This Notebook Belongs to :

AF882644

## Rationalize the Denominator

**Rationalizing the denominator** means the process of moving a root from the denominator to the numerator.

$$\frac{1}{\sqrt{a}+\sqrt{b}} = \frac{1}{\sqrt{a}+\sqrt{b}} \cdot \frac{\sqrt{a}-\sqrt{b}}{\sqrt{a}-\sqrt{b}} = \frac{\sqrt{a}-\sqrt{b}}{(\sqrt{a}+\sqrt{b})(\sqrt{a}-\sqrt{b})} = \frac{\sqrt{a}-\sqrt{b}}{(\sqrt{a})^2-\sqrt{ab}+\sqrt{ab}-(\sqrt{b})^2} = \frac{\sqrt{a}-\sqrt{b}}{(\sqrt{a})^2-(\sqrt{b})^2} = \frac{\sqrt{a}-\sqrt{b}}{a-b}$$

*Example:*

$$\frac{1}{\sqrt{5}+\sqrt{3}} = \frac{1}{\sqrt{5}+\sqrt{3}} \cdot \frac{\sqrt{5}-\sqrt{3}}{\sqrt{5}-\sqrt{3}} = \frac{\sqrt{5}-\sqrt{3}}{(\sqrt{5}+\sqrt{3})(\sqrt{5}-\sqrt{3})} = \frac{\sqrt{5}-\sqrt{3}}{(5)^2-\sqrt{15}+\sqrt{15}-(\sqrt{3})^2} = \frac{\sqrt{5}-\sqrt{3}}{(\sqrt{5})^2-(\sqrt{3})^2} = \frac{5-3}{5-3} = \frac{\sqrt{5}-3}{2}$$

## Algebra Identities

**Algebra identities** are algebraic equations which are valid for all values variables. They are used for factoring polynomials and in the computation of algebraic expressions and solving different polynomials.

$$(x+a)(x+b) = x^2 + (a+b)x + ab$$

$$a^2 - b^2 = (a+b)(a-b)$$

$$(a+b)^2 = a^2 + 2 + b^2$$

$$(a-b)^2 = a^2 - 2ab + b^2$$

$$a^3 - b^3 = (a-b)(a^2 + ab + b^2)$$

$$a^3 + b^3 = (a+b)(a^2 - ab + b^2)$$

## Logarithms

A **logarithm** is an operation that determines how many times a certain number, called the base, is multiplied by itself to reach another number. It's the inverse function to exponentiation (exponents).

$$b^m = a \Rightarrow \log_b a = m$$

$$\log_b 1 = 0$$

$$\log_b b = 1$$

$$\log_b xy = \log_b x + \log_b y$$

$$\log_b \frac{x}{y} = \log_b x - \log_b y$$

$$\log_b x^m = m \log_b x$$

$$\log_b x = \frac{\log x}{\log b}$$

## Composite Functions

A **function** is a rule which relates a given set of inputs to a set of possible outputs **Function notation**, $f(x)$, is read as "$f$ of $x$," where $f$ is the name of the function and $x$ is the input of the function.

A **composite function** is a function written inside another function. Composition of a function is done by substituting one function into another function.

$f[g(x)] = (f \circ g)(x)$ is read as "$f$ of $g$ of $x$." The function $g(x)$ is called the **inner function**, and $f(x)$ is the **outer function**. To solve composite functions, substitute the variable $x$ that is in the outside function with the inner function.

*Example:*

Given the functions $f(x) = 2x + 6$ and $g(x) = x - 1$, find $(f \circ g)(x)$.

Step 1: Substitute $(x - 1)$ for $x$ in $f(x)$.

$(f \circ g)(x) = 2(x - 1) + 6$

Step 2: Simplify.

$(f \circ g)(x) = 2(x - 1) + 6 = 2x - 2 + 6 = 2x + 4$ or $2(x + 2)$

## Permutations and Combinations

**Permutations** are used to find the different arrangement of $r$ things from the $n$ available things, and **combinations** are used to find the different groups of $r$ things from the available $n$ things.

$$n! = n \times (n-1) \times (n-2) \times \cdots 3 \times 2 \times 1$$

$$nCr = \frac{n!}{(n-r)!\,r!}$$

$$nPr = \frac{n!}{(n-r)!}$$